Uno Programming
digitalWrite

Shannon Davis

ebonygeek45@gmail.com

Copyright © 2017 Ebonygeek45
All rights reserved.
ISBN-13: 978-1542495677
ISBN-10: 1542495679

DEDICATION

Emma Davis my loving mother is my rock. Without her I don't know where I would be. I dedicate this book to her. One day she may see it.
Ebonygeek45 aka Shannon Davis

ALSO BY <AUTHOR>

thegeeks.gq: Fresh New Website

Books

Uno Programming digitalWrite: More on C++

Beginners Edition 1

Uno Easy Starter Project : LED Cube Arduino Uno Building and Coding

---

Videos

Ebonygeek45: Youtube Channel(Subscribe To Channel)

https://www.youtube.com/Ebonygeek45

Keep updated and in touch with ebonygeek45

Ebonygeek Speak Arduino: Website

https://sites.google.com/site/ebonygeekspeakarduino/

Visit us of Facebook: Facebook

https://www.facebook.com/profile.php?id=100016873423390

Follow us on Twitter : Twitter

https://twitter.com/ebonygeek/

Email us

ebonynerd45@gmail.com

TABLE OF CONTENTS

DEDICATION _____ iii

ALSO BY AUTHOR _____ iv

INTRODUCTION _____ vi

CHAPTER 1: THE START OF OUR PROJECT _____ 1

CHAPTER 2: EXPLAINING THE CODE MORE _____ 30

CHAPTER 3: VARIABLES _____ 36

CHAPTER 4: CREATING THE SOS PROGRAM ____ 46

CHAPTER 5: FUNCTIONS _____ 57

CHAPTER 6: FUNCTIONS AND VARIABLES _____ 77

CHAPTER 7: MORE PASSING VARIABLES _____ 106

CHAPTER 8: BEHAVIORS AND ATTRIBUTES ____ 118

CHAPTER 9: FOR LOOP _____ 126

CHAPTER 10: IN CLOSING _____ 134

ABOUT THE AUTHOR _____ vii

INTRODUCTIONS

The programming, electronics, prototyping, and the world in general is huge. People live, work, and play in this world. We love the new gadgets and gizmos that come out.

Life is good isn't it?

Children get these toys as a gift or just begged for it until they got it.

That is when the love for new exciting things start. They play with it, bang it around, uh oh it is broken.

That is the life of gadgets and gizmos. From childhood to retirement and beyond we love technology.

Some kids ask why?

So they find a screw driver or in my childhood a butter knife, or even pull it apart. They pull out the stuffing or break apart any casings and find electronics. Some kids toy with it and get a beep or movement.

The minds of the curious are tweaked.

They ask why?

Most of the time they get a lecture or even a spanking or time out.

How could you break that new (and most of the time expensive) toy?

But sometimes, more rarely than most an adult or even another kid can answer to them – WHY.

Those fortunate kids grow up and keep tinkering and learning.

Some adults go through the same process later in life. They are called geeks and nerds and picked on like that is a negative thing. It is not and most of them keep expanding their knowledge and grow.

They are the troubleshooters, problem solvers, and "go to" people of the world.

The workforce of the world reward and acknowledge these fortunate ones. These ones are rarer then most.

For the larger part of the workforce these special ones are underpaid, overworked, demoralized, and expendable.

Maybe even further used and exploited by student lending in the business of college education for a profit.

Opportunist, exploiters, leeches, users, spongers, parasites, social engineers, etc. All come to mind in the real work world.

But they tinker on and that is the strength of these people. They are inventors, engineers, and genius of the day.

Nikola Tesla, Micheal Faraday, Thomas Edison, Benjamin Franklin, Marie Curie, Charles Babbage, Robert Kahn, Vint Cerf, to name a few.

Later on you have the notable of the day like: Bill Gates, Linus Torvalds, Massimo Banzi, Robert Portugal, comes to mind.

There are countless others. The nameless and faceless ones

that we are not aware of. But they are just as notable.

Then you have your common everyday men and women.

They tinker on in basements, garages, at home in laboratory's, workshops, incubators, etc.

They find materials in junk yards, stores, and on the internet.

Toys, computers, and electronics taken apart and reused to make incredible things. They are anywhere from the young to the old. Spanning through the wealthy to the impoverished.

They do it for their own personal fulfillment or gain. To fix something that is broken. To make something for someone as a gift. Or any other small reason they can find. To sell or be the first to come out with an idea people can improve on.

These people are the salt of the earth and the reason we enjoy what we enjoy today.

We need to encourage this. To catch people like this when they are young for special education and training.

Encourage youths and adults to understand and create the technology they so love.

Wouldn't that be a wonderful world?

Imagine how much further we would be.

It has already started. Through Wikipedia, YouTube, Github, and open sourcing. More need to be done to get us there. How fast may very well depend on you and me.

CHAPTER 1: THE START OF OUR PROJECT

Congratulations on starting in the electronics prototyping world. Today is a good day to learn something.

It is always good to start small and build your way up. Get a good foundation in place, then build up from there.

It will be assumed that you have an Uno. That can be an official Arduino, Make, clone, or one you built yourself.

It will also be assumed that you have installed the Arduino IDE. It would be a good idea to install it now if you haven't already. The link is below.

https://www.arduino.cc/en/Main/Software

These are the main things you will need. Without them you will not get the full concept of learning to use them.

For the Uno I would suggest getting one from the store. That way you can return it if it does not work. Ebay is always a good resource for components, but in my experience the Uno's did not work (sorry). It was not worth the money to send them back.

Back to what I was saying. There is nothing wrong with reading through this book first. Project 1 is strictly for beginners.

Beginners in prototyping and programming. That is what you will be doing. The project is for the young through the old and

Uno Programming digitalwrite   Ebonygeek45

if you can read that is the only real requirement.

This book works on giving a programming foundation to ease you through the beginning stages.

Getting Started

To get the full experience of the project you will need:

1. UNO (Arduino, Make, Inland, Clone, or one you built)

2. Breadboard

3. 3 LED's (One will be used for our project)

4. 3 220 (ohm) Resisters

5. Jumper wires (20 gauge or 22 gauge wire) or pre-made.

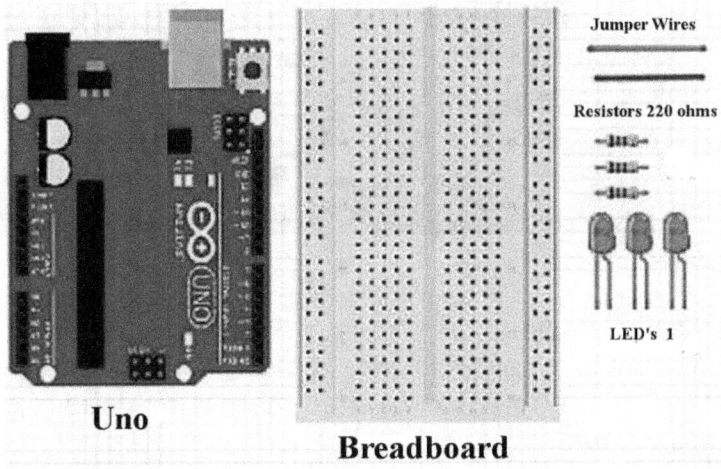

Arduino IDE (Installed)

Uno Programming digitalwrite

```
1  void setup() {
2    // put your setup code here, to run c
3
4  }
5
6  void loop() {
7    // put your main code here, to run re
8
9  }
```

This list is simple and inexpensive as possible. The Arduino IDE is simple to install. There is much information on the internet as to how to install it, so there is no need to go through it here.

So let's dive right in.

Uno Programming digitalwrite   Ebonygeek45

Our Project

That was quick wasn't it. So lets get to it.

The first thing you will need to learn is how to get that led blinking.

That is the purpose of the Uno.

The Uno is a micro controller. It is a small computer that allows you to make electronics do what you want them to. There are micro controllers of some sort everywhere running all kinds of things. As you learn more, you will have a better understanding how your everyday electronics work.

You have everything on the list. You're ready to go. Perhaps you have already done the blink program.

You ask "What is the first project?"

Answer: It is creating an SOS Morse Code Blinker.

Uno Programming digitalwrite — Ebonygeek45

You say "A what??"

I am not surprised. Most kids of the day with cell phones and tablets have never even heard of Morse code. Even if you have heard of it you would never use it right?

Back in the day when there were no phones, there was Morse code. It was a way to send messages through an electrical telegraph system. We won't be doing that, however interesting it may sound.

We will be programming the led to blink SOS in Morse code.

Pin Out For Circuit

Let's start with the blink program.

This will be the start of our Morse blinker.

Tip: Before you put miles on your Uno stop for a moment. If you are like me, your Uno did not have any protection under the bottom.

If you can find rubber feet like the one you see on lamps and other things, get them. I got mine from Home Depot. Stick them to the bottom of the Uno. If you can't find those use electric tape, or something to protect the bottom of the Uno.

Uno Programming digitalwrite   Ebonygeek45

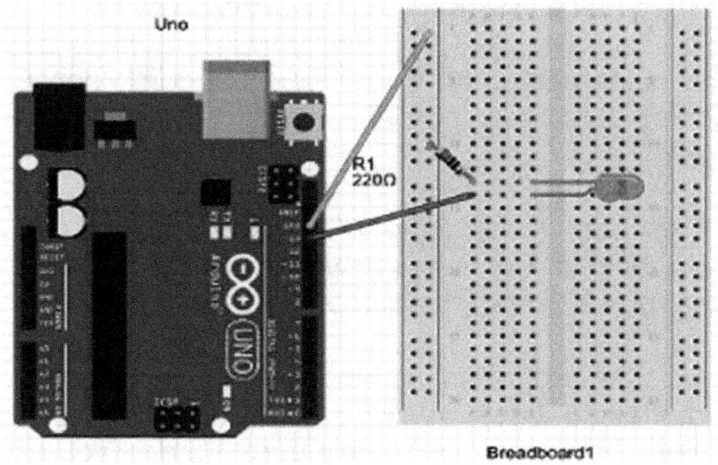

Take a good look at your Uno board. The picture above shows how you want to wire your circuit.

+ Step 1: Start with the GND (aka Ground) over pin 13 on your Uno, as you see from the picture. The ground goes down the row you connect it to on the left and the right. If you put it on the blue side you will know it is ground (less confusion).

+ Step 2: Connect your resistor from the ground row to the ground leg of the led (shortest). The ground leg of the resistor is the cathode.

+ Step 3: Connect pin 13 on the Uno to the positive leg of the led (longest). The positive leg is the anode.

This is a very simple circuit. We will use it to show and verify that our code examples are working.

Programming in C++

Uno Programming digitalwrite   Ebonygeek45

If you know C++, great. You're good to go.

This is the perfect book for those that don't know C++. Are you looking at the Arduino IDE thinking, I don't know what to do now?

Click around the Arduino IDE. You are not going to know what everything is. But that is how you start to get a feel for it. It is a good place to start. I still use it with Notepad++.

C++ is the programming language you would use with the Arduino IDE. Don't worry because we are going to go over it in detail. That is the point of this book. To give you a good C++ programming foundation to build upon.

The Arduino IDE has a number of examples to start you off. So pull up your Arduino IDE if you don't have it up already.

A Quick Look At the Arduino IDE

Drop Down Menu – Use to find all the menu's provided to you.

Button Ribbon – For quick access to the more commonly used selections in the menu.

Start Up Code – Is shown here in the Editor section. This is where you will do your C++ code. The code pictured is providing you with the two required functions for every sketch. (Will explain shortly.)

Generic Code – The tab within the white area is showing an example of the name that is given to every sketch started. It is strongly recommended to change the name to something more descriptive. That way you will know what the sketch is about on sight.

Underneath it all is a black window that you will be paying a lot of attention to. It will tell you if your code is correct. It will tell you if your code has errors. Like most editors, you will have to research the errors to see how to correct them. This is a big part of coding.

Uno Programming digitalwrite   Ebonygeek45

Again, this book will give you a lot of examples in coding to help you avoid most of the newbie errors that stumps us.

Let's take a look at the buttons on the button ribbon.

When you hover over the button it will tell you what it is.

Verify   -   Checks the code and tell you if there are errors.

   It is advised to verify your code often. Verifying often will help you correct errors as you go along. It is harder to find errors if you do not verify your code until you're finished. Be careful, some errors are a case of you have done something but not completed it.

Upload   -   Sends your code to your board. In this case the Uno.

   Upload will also show any errors in the code. It tells you if your Uno is not connected(As in you need to plug the USB cord into the Uno). Also, it will tell you if you are using the correct port for the Uno (Each Uno will have it's own port number.).

New - For when you want to start a new sketch. It will open in another window.

Open - You may have many sketches.

If you want to open one of them you could click this button to open that file. It will also open in its own window.

Save - Just like verify, it is a good idea to save often.

It would be very frustrating to code a long program that is

lost. Take my advise and verify and save often. It will save you a lot of hassle.

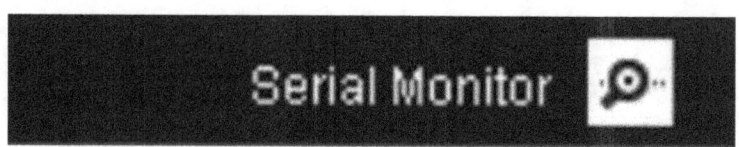

Serial Monitor - Serial Monitor is a way of communicating with the Uno.

You can run code that can print out to the serial monitor. This can have a lot of uses. You can even troubleshoot your code with it.

These selections on the button ribbon are the most used. They are there for your convenience in an attempt to make your coding easier.

A big "Thank you" to the folks who designed the Arduino IDE. Believe me it is better than having to "make" or "build" your programs from scratch.

The Blink Example Code

Going further into the Arduino IDE. We are going to take a look at the example blink. We will go through the Drop Down Menu's. We should be very familiar with it from the many programs that have it.

File > Examples > 01. Basics > Blink

Uno Programming digitalwrite   Ebonygeek45

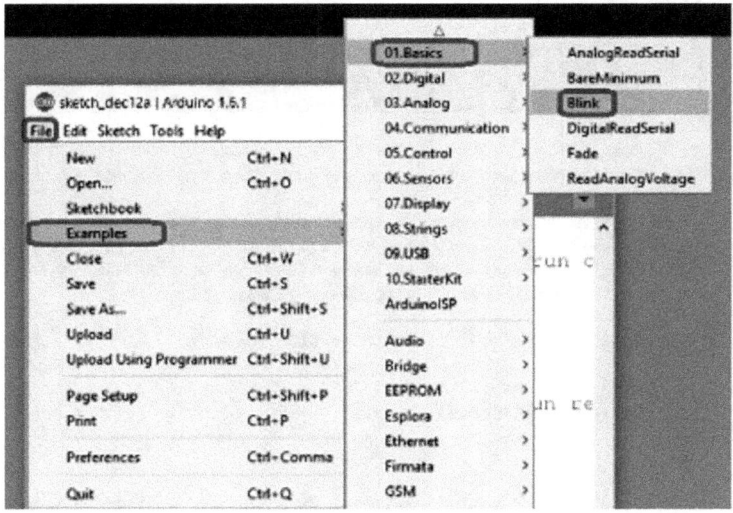

As you can see there are a lot of examples. Of course you can use them as you need them. We are going to need Blink.

Another one that you may find convenient is Bare Minimum.

It brings up the required code for every sketch. If your Editor comes up blank, go to Bare Minimum.

But for now go ahead and click on Blink.

```
 1 /*
 2   Blink
 3   Turns on an LED on for one second, then off for one second, repeatedly.
 4
 5   Most Arduinos have an on-board LED you can control. On the Uno and
 6   Leonardo, it is attached to digital pin 13. If you're unsure what
 7   pin the on-board LED is connected to on your Arduino model, check
 8   the documentation at http://www.arduino.cc
 9
10   This example code is in the public domain.
11
12   modified 8 May 2014
13   by Scott Fitzgerald
14  */
15
16
17 // the setup function runs once when you press reset or power the board
18 void setup() {
```

Comments

The first thing in the blink program you will see is comments from its creator. Sketches in the examples will have these comments. It gives you information about what the sketch does.

Comments are very important in your coding. You should use them. It is a sign of a good programmer. Comments help you and even others using your code to understand what you are doing quickly. Most sketches you get on the internet will also have comments.

// This is a single line comment.

/*

Uno Programming digitalwrite   Ebonygeek45

    This is a multiple line comment

    You can add as many lines as needed.

*/

It is a good idea to pay attention to the comments you and others add in code.

The Blink code

****

/*

Blink

Turns on an LED on for one second, then off for one second, repeatedly.

Most Arduinos have an on-board LED you can control. On the UNO, MEGA and ZERO it is attached to digital pin 13, on MKR1000 on pin 6. LED_BUILTIN is set to the correct LED pin independent of which board is used.

If you want to know what pin the on-board LED is connected to on your Arduino model, check the Technical Specs of your board at https://www.arduino.cc/en/Main/Products

This example code is in the public domain.

modified 8 May 2014

by Scott Fitzgerald

Uno Programming digitalwrite   Ebonygeek45

modified 2 Sep 2016

by Arturo Guadalupi

modified 8 Sep 2016

by Colby Newman

*/

// The setup function runs once when you press reset or power the board

```
void setup()
{
    // Initialize digital pin LED_BUILTIN as an output.
    pinMode(LED_BUILTIN, OUTPUT);
}
```

// the loop function runs over and over again forever

```
void loop()
{
    // turn the LED on (HIGH is the voltage level)
```

Uno Programming digitalwrite					Ebonygeek45

```
    digitalWrite(LED_BUILTIN, HIGH);

    // wait for a second

    delay(1000);

    // turn the LED off by making the voltage LOW

    digitalWrite(LED_BUILTIN, LOW);

    // wait for a second

    delay(1000);
}
```

****

Next we come to the actual coding for Blink shown above,

You see the many comments given by the author of this code.

See how helpful it is.

You are encouraged to make detailed notes. The Arduino IDE offers you a way to manage them. Kind of fold them out of your way when you are developing your code.

Go to your drop down menu's.

File > Preferences

Uno Programming digitalwrite

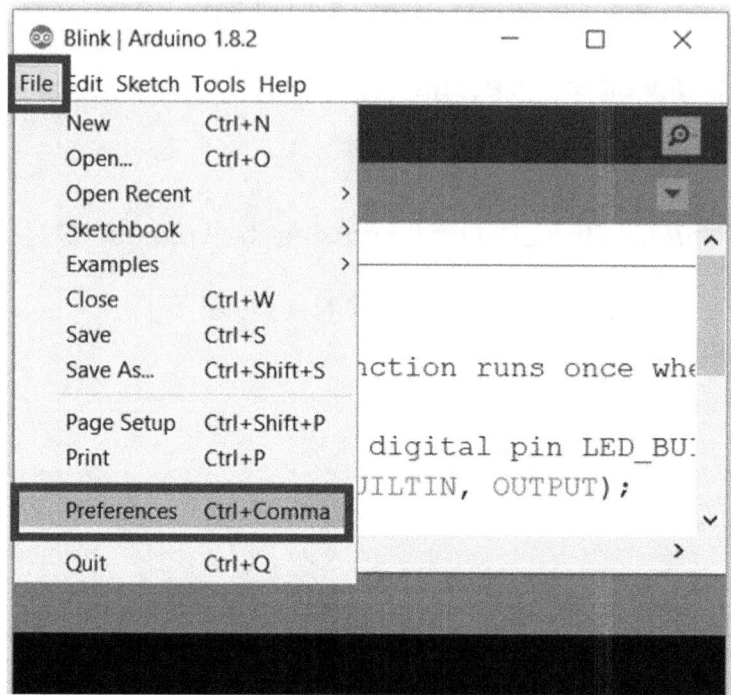

Check the option "Enable Code Folding" > "OK"

You can also check "Display Line Numbers"

Make sure to click:"OK" for the preferences to save.

Preferences

Sketchbook location:
C:\Users\Shannon\Documents\Arduino

Editor language: System Default   (re

Editor font size: 16

Show verbose output during: ☐ compilation ☑ upload

Compiler warnings: None

☑ Display line numbers
☑ Enable Code Folding
☑ Verify code after upload
☐ Use external editor
☑ Check for updates on startup
☑ Update sketch files to new extension on save (.pde -> .ino)
☑ Save when verifying or uploading

Additional Boards Manager URLs:

More preferences can be edited directly in the file
C:\Users\Shannon\AppData\Roaming\Arduino15\preferences.txt

Go back to the editor. You will notice for the multiple line comments you can click the "+" and "-" to expand or collapse them.

Uno Programming digitalwrite

```
sketch_dec12b §
/*
7  const int pin13 = 13;
8  const int _1second = 1000;
9
10 // setup function runs once when you press reset or po
11 void setup()
12 {
17
18    // the loop function runs over and over again foreve
19 void loop()
20 {
21    // digitalWrite(13, HIGH);
```

That also applies to the functions. That is very convenient for many reasons.

Looking At The Blink Code

This is how my code comes up for blink in the Arduino IDE.

Uno Programming digitalwrite

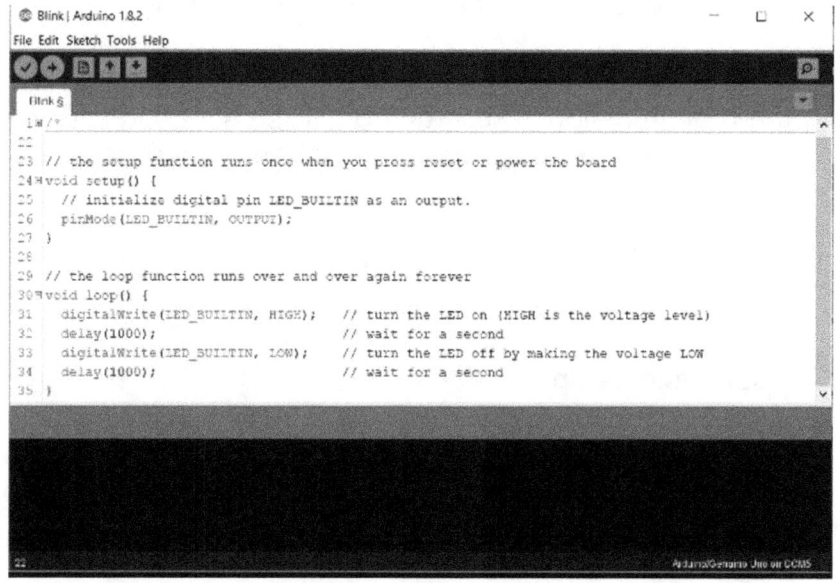

The first thing I do is go through it and edit how I like my code to look.

Readability in your code is very important. A tab here, a space there, add a line. After setting it up the way I like we come up with this.

If you find that it's not doing what you want it to do. Just close it out and reopen it. Sometimes it likes to be a little stubborn.

The code above is well commented.

What are you looking at? You ask.

It is C++. If you are new to it, it may look strange to your eyes.

Uno Programming digitalwrite

void setup and void loop are both functions.

All C++ code makes use of functions. Normally you would see the main function. The Arduino IDE sets it up in their own way. Think of void loop as main. It is where your functions will be called or coded.

It is called void loop because it will loop through your code over and over again. We will go into loops a bit later.

About void setup:

It is showing that it is initializing LED_BUILTIN. It is doing exactly what it says. It sets up the pins you will use on the Uno.

You would not put this code in the loop. It only needs to run once at the startup of your program.

Inside the Curly brackets:

{ }

Are the instructions for the Blink code for both the setup function and the loop function. It is what we want the led to do. This is the strength of the Uno Micro Controller. You can make electronics do what you want them to do.

In this case the instructions are to have the led blink for I second.

So for this example, you should have your pin out done. We now have the code sample. Go ahead and click your verify button.

Uno Programming digitalwrite											Ebonygeek45

You should notice the results of the code running in the black section below the editor.

Your code should verify and you should see the results above. Remember, you are not making any changes yet.

Go ahead and click Upload.

This uploads the code into the Uno through the USB cord.

You should see the same kind of results as when you verified, or something like it. If so you are good to continue on. Depending on which version of the Arduino IDE you have the results screen may look different. The sketch may even be a little different. It should all work the same.

There should be no errors at this point.

Problems with Port and Com?

Uno Programming digitalwrite   Ebonygeek45

But...

What if you get an error that you're using the wrong port for your Uno.

How do you know which port to use?

You have to do some troubleshooting.

Because we may be on different systems on computers, I will give an example for computers running Windows 10.

We are looking for the Device Manager. If you're on another type of system you would need to find it.

Go to your start menu for your computer. You know the Microsoft button at the bottom corner of your computer.

There may be a search box next to it or when you click on it.

In that search bar you would type "Device Manager.

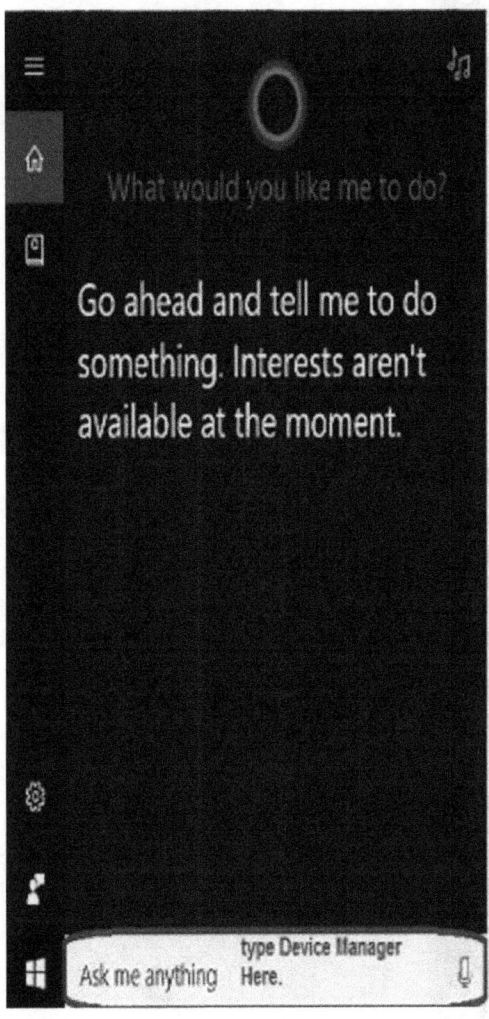

It will bring it up and you would click on it.

Uno Programming digitalwrite

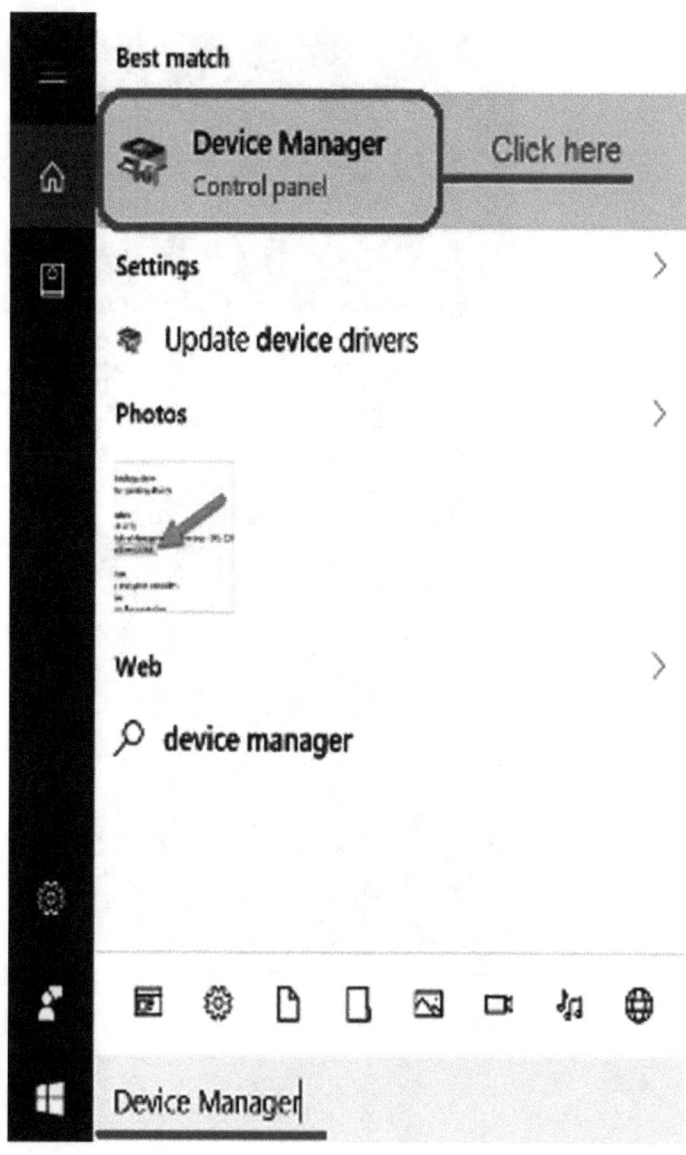

Uno Programming digitalwrite   Ebonygeek45

Scroll down to the Ports as shown here.

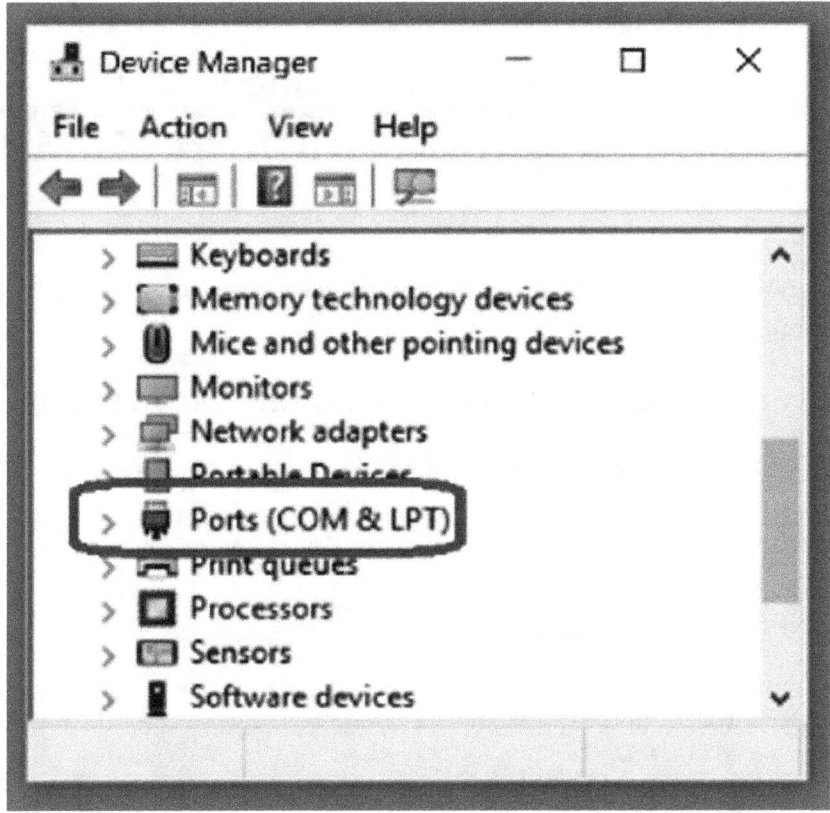

Click on the arrow.

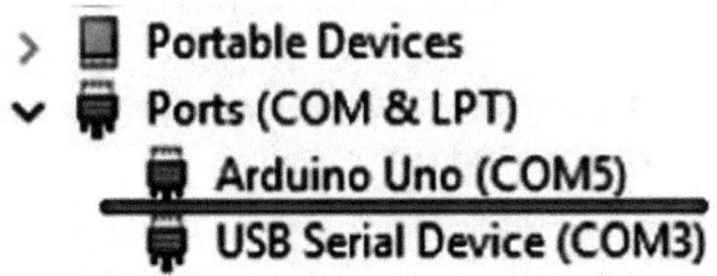

When you click on the arrow, it will show you the ports you have in use.

For example, it is showing me my Uno is on (COM5).

Go back to your Arduino IDE. You want to go to:

   Tools > Port

From your Drop Down Menu.

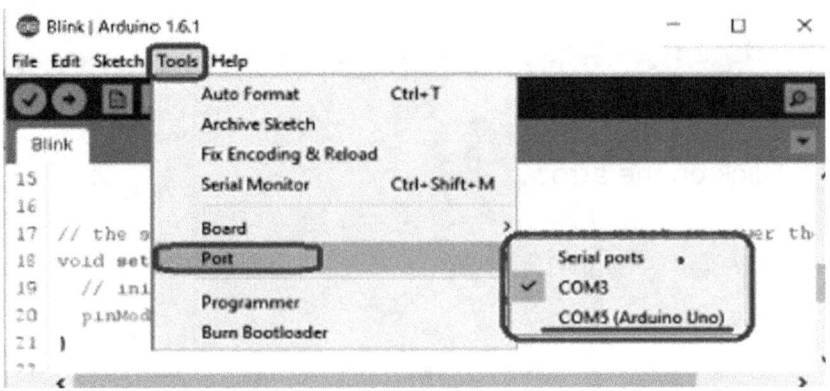

Here it is showing that COM3 is selected. But we know it should be COM5. Select the correct port by clicking on it. Then click upload and everything should be fine. Keep in mind yours may be different. Use the COM information showing on your computer.

Observe your led.

    Is it blinking for 1 second?

If not, go back through the steps again until it is working.

Otherwise, congrats.

Summary

We should all be on the same page now.

You uploaded a program into the Uno Micro Controller to control the led.

At this point you should have:

> The Uno hooked up correctly to the circuit on the breadboard.

> The Uno connected to the computer.

> The blink sketch should have verified and uploaded.

> The led blinking for 1 second.

You're on your way.

Uno Programming digitalwrite

Keep following along.

CHAPTER 2: EXPLAINING THE CODE MORE

At this point everything should be up and running correctly. The led should be blinking away. Now we can focus on some basics of C++.

Remember led means light emitting diode.

Starter Functions And Syntax

Let's backtrack a little. I know good comments are there. So keep and eye on them. A little more explanation needs to be given.

We already spoke about the setup and loop function. But we didn't go into what was within the setup and loop function.

The Arduino IDE people created (freebie) functions and code for you to use. Again in an attempt to help you with your coding. This makes good starter code for you.

In the setup function, we have one function.

    pinMode

        Is a function that shows which pin you are using.

        Then set it to either OUTPUT or INPUT.

        It will go out to the Uno so for this example we are using OUTPUT.

Uno Programming digitalwrite   Ebonygeek45

The syntax for pinMode is;

   pinMode(<pin used>, <OUTPUT or INPUT>);

What does that mean?

pinMode is the function used.

The () after it is it's parameters.

The ; after the parameters is a semi colon and how you end a line of code in C++.

Inside the parameters are my instructions to attempt to make it easier for you to understand when explaining syntax.

In this book whenever you see something in <> angle brackets, it means to replace whatever is in the <> angle brackets with your own code IE the pin used, OUTPUT or INPUTis required code:

For example; Either OUTPUT or INPUT.

So if I give the syntax:

   pinMode(<pin used>, <OUTPUT or INPUT>);

It explains what is needed for the pinMode function.

   pinMode(LED_BUILTIN, OUTPUT);

Or if you have an older Arduino IDE you may see...

   pinMode(13, OUTPUT);

Uno Programming digitalwrite							Ebonygeek45

No worries, both codes are doing the same thing.

<pin used> is replaced with either LED_BUILTIN or 13 because that is the digital pin used on the Uno.

<OUTPUT or INPUT> is replaced with OUTPUT because that is the mode we want to use.

To continue on to the loop. We see 2 more functions.

digitalWrite

Writes to the digital pin and/or pins either HIGH or LOW. That is what we are dealing with in this book.

HIGH - meaning on, over 0, true, yes positive or simply HIGH. However you want to think about it.

Low meaning - off, 0, false, no, negative or simply LOW. Again, however you want to think about it.

The syntax for digitalWrite is;

digitalWrite(<pin used>, <HIGH or LOW>);

delay

Pause the program for the amount of time in the parameters(by milliseconds). 1000 milliseconds is a second.

The syntax for delay is;

Uno Programming digitalwrite   Ebonygeek45

delay(<milliseconds>);

That is the 3 ready made functions we will be using, and examples of my helper syntax.

Now Take a look at our code.

The Blink Code

****

// Blink

// the setup function runs once when you press reset or power the board

void setup()

{

    // initialize digital pin LED_BUILTIN as an output.

    pinMode(LED_BUILTIN, OUTPUT);

}

// the loop function runs over and over again forever

void loop()

{

    // turn the LED on (HIGH is the voltage level)

```
digitalWrite(LED_BUILTIN, HIGH);
// wait for a second
delay(1000);
// turn the LED off by making the voltage LOW
digitalWrite(LED_BUILTIN, LOW);
// wait for a second
delay(1000);
}
```

****

We are going to start adding code to optimize it.

Optimize

Rearrange or rewrite (data, software, etc) to improve efficiency of retrieval or processing.

We would ask ourselves:

Is there anything we see repeated?

Keep that question in mind and continue on to the next chapter.

Summary

Uno Programming digitalwrite   Ebonygeek45

Yes, I know that it was barely a Chapter.

It is important to understand it though.

Go over your code in your IDE and the information about the functions and more importantly, my syntax.

Once you feel like you understand it continue on.

CHAPTER 3: VARIABLES

Remember our question about starting to optimize our code?

Is there anything we see repeating?

Yes, LED_BUILTIN(or 13) and 1000.

Using Variables

How do you optimize the code for those numbers which are values?

Variables would be the answer.

Variables are a way of holding a value by its type.

There are many types you can choose from.

For now we are going to use int for integer.

Why use Variables?

They allow you to only have to change a value once if you need to. Rather than going through lines of code changing the same values.

This may seem simple to you to change now as a value.

Some programs have hundreds and thousands of lines of coding. In those cases you want as many advantages as you can get.

Uno Programming digitalwrite   Ebonygeek45

Also, in going further into what you can do with C++ you need to use variables.

We will add the variables over the setup function.

We do not want to put them inside the setup or loop function. That would limit them to whatever function they were added to.

We already know the variable types will be int.

We know the values will be 13, and 1000.

All we have to do is name them something that would describe what they are.

13 is for the pin number. Let's name it pin13.

1000 is for the delay in milliseconds. 1000 milliseconds is a second.

My choices are _1second or second1.

You can use any name you want, just so the name makes sense to you.

As a rule: You can't start a variable name with a number, but you can start it with an underscore then the number.

So my variable names will be pin13 and _1second.

Syntax for variables:

Uno Programming digitalwrite

<type> <name> = <value>;

Using the syntax example above we make our new variables.

int pin13 = 13;

int _1second = 1000;

```
15
16  int pin13 = 13;
17  int _1second = 1000;
18
19  // the setup function runs once v
20  void setup()
21  {
22      // initialize digital pin 13 as
23      pinMode(13, OUTPUT);
24  }
```

We add that over the setup function. Then verify the code.

Did your code verify?

If not, try another name for the variable. Remember, it can't start with a number. Also, it can't be a keyword.

My choice of variable names above runs with no problem.

Now we have our variables. But we need to add something else to them.

The variables we have created now are regular variables.

Regular variables values can change.

Constant Variables

There are "Constant" variables too. It is a variable that does not change.

To clarify. We can change it all we want in the editor. But it will not change once it is uploaded to the Uno.

The pin will stay at 13 (or whatever pin you want to use, as long as it is on the digital side of the Uno).

The delay will be for however many milliseconds we put in its parameters. You can think of Constant variables as Read Only variables.

Syntax for Constant variables.

```
const <type> <name> = <value>;
```

```
15
16  const int pin13 = 13;
17  const int _1second = 1000;
18
19  // the setup function runs on(
20  void setup()
21  {
22      // initialize digital pin 1:
```

We simply add const to the beginning of the variables. Then verify the code.

You have the variables created.

Let's replace 13 with our variable pin13 in the setup and loop functions.

The code is below how it now appears and verifies.

The Blink Code

****

// Blink.ino

/*

   Replacing values for pinMode and digitalWrite with constant

Uno Programming digitalwrite   Ebonygeek45

variables.

Notice I did not delete the original code. I just commented them out and added the new code below them.

That's a way to test code, but keep the original code until you know the new code is working.

*/

// New Variables

const int pin13 = 13;

const int _1second = 1000;

/* setup function runs once when you press reset or power the board. */

void setup()

{

   // initialize digital pin 13 as an output.

   // pinMode(LED_BUILTIN, OUTPUT);

   pinMode(pin13, OUTPUT);

}

// the loop function runs over and over again forever

```
void loop()
{
    // digitalWrite(LED_BUILTIN, HIGH);
    // turn the LED on (HIGH is the voltage level)
    digitalWrite(pin13, HIGH);
    // wait for a second
    delay(1000);

    // digitalWrite( LED_BUILTIN LOW);
    // turn the LED off ( LOW is the voltage level)
    digitalWrite(pin13, LOW);
    // wait for a second
    delay(1000);
}
```

****

Verify the code.

Now that is done, go ahead and do the same thing for the delay variable.

Uno Programming digitalwrite			Ebonygeek45

Verify the code and upload it.

What were your results??

Summary:

Your results should have been exactly the same for the variables as it was for the numbers (values).

If it wasn't take a good look at your code to see if you did something wrong.

Hint: Check that semicolons aren't missing at the end of lines of code. Make sure the names are correct.

> Make sure that the code is correct and you didn't make any mistakes with the syntax.

Your code should look like the code below.

The Blink Code

****

```
// Blink.ino

// New pin variable

const int pin13 = 13;

// New delay variable
```

Uno Programming digitalwrite

```
const int _1second = 1000;

/* setup function runs once when you press reset or power the board. */

void setup()
{
    // initialize digital pin 13 as an output.
    pinMode(pin13, OUTPUT);
}


// the loop function runs over and over again forever
void loop()
{
    // turn the LED on (HIGH is the voltage level)
    digitalWrite(pin13, HIGH);
    // wait for a second
    delay(_1second);
```

// turn the LED off (LOW is the voltage level)

digitalWrite(pin13, LOW);

// wait for a second

delay(_1second);

}

****

That is the way variables are used. They are very handy. We will be doing more with them later.

CHAPTER 4: CREATING THE SOS PROGRAM

Nervous??

Don't be. We are going to take it slow in this chapter.

It may be too easy for you because it won't be anything we have not done in the past chapters.

But it's good practice.

So whatever you do, do not copy and paste the code.

Get the extra practice in.

Algorithms

You may have heard the word "Algorithms" before, but do you know what it means?

Just using the word make me feel more intelligent.

> Algorithm
>> A process or set of rules to be followed in calculations or other problem solving operations, especially by a computer.

That's it, so simple.

So let's make a simple algorithm for our program.

Uno Programming digitalwrite   Ebonygeek45

Why? You say.

Good planning, less hair pulling, less aggravation and confusion.

I used to think it was a waste of time when I could just get to the coding. Getting lost in what I was trying to do all the time taught me a hard lesson. One that I advise you to avoid.

It can be very simple and can be done with your comments.

How? You say.

    1. Think out what you want to do when programming,

        1. When you do this there is less of a chance of you getting lost in what you are trying to do.

        2. Less time is wasted figuring out what to do next.

    2. Think it out, then write it out like below.

        1. The SOS program depends on "." and "-".

        2. We will just refer to it as "dot" and "dash".

        3. The dot is a short blink.

        4. The dash is a long blink.

        5. We will show this as seconds to start off slow.

        6. Dot will be 1 second and the dash will be 3 seconds.

Uno Programming digitalwrite　　　　　　　　Ebonygeek45

　　7. S is 3 dots in Morse code, and O is 3 dashes.

　　3. That is how we will make our light blink SOS.

　　4. Heavy sigh and have a cookie. Chocolate Chip with Pecans sound good.

Now you know what you need to do. So plot it out in the Editor.

Below is my example of the very simple SOS code plotted out.

Uno Programming digitalwrite

```
B3digitalWriteCreatingTheProgram1a §
28 {
29    /* This is gonna be a bit of coding,
33    /// S
34    // .
35    // .
36    // .
37
38    /// O
39    // -
40    // -
41    // -
42
43    /// S
44    // .
45    // .
46    // .
47 }
```

As you can see the Algorithm is very simple.

A lot better than struggling through it trying to figure out what to do "on the fly".

You could probably figure out the code for it just by this Algorithm.

Feeling Froggy??

Uno Programming digitalwrite   Ebonygeek45

Feel Froggy??   Then jump.

See if you can do it on your own with the above Algorithm. Then come back here and compare your code with what we have here.

If you make mistakes, try to work through them. If it is too hard come on back and we will continue on. But working through the algorithm and any errors will help your "trouble-solving" skills develop nicely. Take your time.

If you don't think your ready that is fine too. That is what we're here for.

That's a plan.

Roll Up Your Sleeves

Did you figure it out?

If so, good job,

If you got stumped... Well, that happens (a lot, frankly at the beginning).

So lets program what we have listed.

The first dot for S is our blink code, is it not?

Yes, it is.

Here is one way to code For S

// S

// .

// Turns digital pin on HIGH

digitalWrite(pin13, HIGH);

// Pause it on high for 1 second.

delay(_1second);

// Turns digital pin off LOW

digitalWrite(pin13, LOW);

// Pause it off LOW for 1 second.

delay(_1second);

OK, so let's repeat it 2 more times for the 3 dots for S. It should be like below.

```
36    /// S
37    // .
38    digitalWrite(pin13,HIGH);
39    delay(_1second);
40    digitalWrite(pin13, LOW);
41    delay(_1second);
42    // .
43    digitalWrite(pin13,HIGH);
44    delay(_1second);
45    digitalWrite(pin13, LOW);
46    delay(_1second);
47    // .
48    digitalWrite(pin13,HIGH);
49    delay(_1second);
50    digitalWrite(pin13, LOW);
51    delay(_1second);
```

You got a free pass on the S.

S should give you an idea on what you need to do for O.

Uno Programming digitalwrite   Ebonygeek45

If you think about it. There should only be one thing that changes for O.

That should be the delay right. It's supposed to be 3 seconds.

How do we change the seconds for O?

We need to add another variable. A Constant Variable.

Remember the syntax:

   const <type> <name> = <value>;

The new variable would be:

   const int _3second = 3000;

We would add it right under the other variables.

```
20   const int pin13 = 13;
21   const int _1second = 1000;
22   const int _3second = 3000;
```

Are you beginning to see the pattern?

Now all we have to do is type out the code for O. Use our new 3 second variable in delay.

Uno Programming digitalwrite   Ebonygeek45

O is shown here for three dashes repeating.

```
57
58    /// O
59    // -
60    digitalWrite(pin13,HIGH);
61    delay(_3second);
62    digitalWrite(pin13, LOW);
63    delay(_3second);
64    // -
65    digitalWrite(pin13,HIGH);
66    delay(_3second);
67    digitalWrite(pin13, LOW);
68    delay(_3second);
69    // -
70    digitalWrite(pin13,HIGH);
71    delay(_3second);
72    digitalWrite(pin13, LOW);
73    delay(_3second);
```

It is the blink code repeating 3 times.

Instead of the delay being at 1 second. It is at 3 seconds.

Uno Programming digitalwrite

You should have been able to knock out the remaining S right away. It is the same as the first S.

Did you remember to verify and save often?

When you uploaded your code after figuring out the SOS code. What happened?

Did the blinks kind of run together?

You don't want them to run together. That is what is happening.

So we need to add another variable.

It is another delay variable.

This one will be added at the end of each letter.

We will name it:

   _4sbetweenLetters.

Remember the syntax:

   const <type> <name> = <value>;

The new Variable:

   const int _4sbetweenLetters = 4000;

Go ahead and add it. Then figure out where to put it in the code.

Summary

How many cookies did you eat?

Good for you.

We now have our program for the SOS blinks.

It wasn't so hard was it?

The code is complete as is. But it can be optimized. I am sure you can see repeating code.

We can also improve the code by adding some communication with the serial monitor. We will do that in the next book.

That's right, we are not done yet. There is more C++ we can learn.

Meet you in the next chapter.

CHAPTER 5: FUNCTIONS

We just worked through an algorithm to create the program.

It optimized our code with the use of variables.

The program creation algorithm is done.

Puff out that chest, and congrats. We are on our way, but we are not done yet.

Catching Up

So where were we?

You had the task of adding the delay between the letters. That would complete the blink part of the program.

How did you do?

To catch up with what our SOS program should be at this point, the code is below.

The SOS code
****

// morseSos.ino

/*

   This is going to be a bit of coding,

Uno Programming digitalwrite

Make use of the algorithm to plot out the process of creating the program.

*/

// Variable Block

// Stores pin number

const int pin13 = 13;

// Stores 1 second value

const int _1second = 1000;

// Stores 3 second value

const int _3second = 3000;

// Stores 4 sec value

const int _4sbetweenLetters = 4000;

/* setup function runs once when you press reset or power the board. */

void setup()

{

Uno Programming digitalwrite

```
    // initialize digital pin 13 as an output.
    pinMode(pin13, OUTPUT);
}
// the loop function runs over and over again forever
void loop()
{
    /// S
    // .
    // turn the LED on (HIGH is the voltage level)
    digitalWrite(pin13, HIGH);
    // wait for a second
    delay(_1second);
    // turn the LED off ( LOW is the voltage level)
    digitalWrite(pin13, LOW);
    // wait for a second
    delay(_1second);
    // .
```

Uno Programming digitalwrite

// turn the LED on (HIGH is the voltage level)

digitalWrite(pin13, HIGH);

// wait for a second

delay(_1second);

// turn the LED off (LOW is the voltage level)

digitalWrite(pin13, LOW);

// wait for a second

delay(_1second);

// .

// turn the LED on (HIGH is the voltage level)

digitalWrite(pin13, HIGH);

// wait for a second

delay(_1second);

// turn the LED off(LOW is the voltage level)

digitalWrite(pin13, LOW);

// wait for a second

delay(_1second);

```
// wait 4 seconds between letters
delay(_4sbetweenLetters);


/// O
// -
// turn the LED on (HIGH is the voltage level)
digitalWrite(pin13, HIGH);
// wait for 3 seconds
delay(_3second);
// turn the LED off( LOW is the voltage level)
digitalWrite(pin13, LOW);
// wait for 3 seconds
delay(_3second);
// -
// turn the LED on (HIGH is the voltage level)
digitalWrite(pin13, HIGH);
// wait for 3 seconds
```

delay(_3second);

// turn the LED off (LOW is the voltage level)

digitalWrite(pin13, LOW);

// wait for 3 seconds

delay(_3second);

// -

// turn the LED on (HIGH is the voltage level)

digitalWrite(pin13, HIGH);

// wait for 3 seconds

delay(_3second);

// turn the LED off (LOW is the voltage level)

digitalWrite(pin13, LOW);

// wait for 3 seconds

delay(_3second);

// wait 4 seconds between letters

delay(_4sbetweenLetters);

Uno Programming digitalwrite

/// S

// .

// turn the LED on (HIGH is the voltage level)

digitalWrite(pin13, HIGH);

// wait for a second

delay(_1second);

// turn the LED off (LOW is the voltage level)

digitalWrite(pin13, LOW);

// wait for a second

delay(_1second);

// .

// turn the LED on (HIGH is the voltage level)

digitalWrite(pin13, HIGH);

// wait for a second

delay(_1second);

// turn the LED off (LOW is the voltage level)

digitalWrite(pin13, LOW);

```
// wait for a second
delay(_1second);
// .
// turn the LED on (HIGH is the voltage level)
digitalWrite(pin13, HIGH);
// wait for a second
delay(_1second);
// turn the LED off ( LOW is the voltage level)
digitalWrite(pin13, LOW);
// wait for a second
delay(_1second);
// wait 4 seconds between letters
delay(_4sbetweenLetters);
}
```
****

Now we should be all caught up.
Working With Functions

Uno Programming digitalwrite   Ebonygeek45

Take a look at your code.

It's a lot of code, isn't it?

Welcome to the world of programming.

But, we got the led flashing SOS. Nice!

The program can be optimized though.

Before we optimized the code with variables.

Now we are going to optimize the code with functions.

Guess what?

We have been working within functions already.

Remember the setup and the loop, we have been working with in our functions.

Take a look at our setup function. It will be easiest because it doesn't have that much code inside it.

It has a:

   <type>

   <name>

   <parameters>

   <curly brackets>

Uno Programming digitalwrite   Ebonygeek45

The syntax is:

   <type> <name>( ) { Instructions for the function }

The setup was:

       void setup() { }

The loop was:

       void loop() { }

For the sake of readability, functions are spread out. What we are used to seeing is.

       <type> <name>( )

       {

          // Instructions for the function

          <code>

       }

We can make our own functions too.

Consider our code. We have 3 blocks of code for the SOS blinks.

O is one block and would be OK. S is the exact same block of code twice.

It would be an improvement to the code to create functions for s and o.

Can you feel an algorithm coming on?

This algorithm is easy because it is an addition to the program.

What do we want to do?:

    // Create function for S

    // Create function for O

That is good enough for now.

You don't have to try to take on everything at once. As you can see we are taking it little by little. Then building the code up.

Functions In Detail

Functions have types just like variables.

Wouldn't the functions be the same type as the setup and the loop?

What type should the functions be?

Those are good questions. Your program gives you hints.

The Arduino IDE is set up so you can "call" your functions in its functions. That is what makes it easier for you to work with it. That is why they are typed void.

What does call mean?

Uno Programming digitalwrite   Ebonygeek45

The best way to explain "call" would be "to use". So to call the function would be to use the function.

When a program is ran in C++ it has to have a starting point. It "call" a function. When dealing with C++ the function it always call first is main. Main is the driver of C++ programs. From there it will call any other functions that you may have in main and run whatever code is within them.

Calling a function transfers the program control from the main to the function it called. It follows the instructions(code) in the function. Then returns control back to the main at the point the function was called. Then continue from that line of code.

For the Arduino IDE the loop is the function that would call our functions. That is how we can use it's functions and our functions without any errors. That is as long as we are writing correct code and using its code correctly. Its main is within the Arduino IDE's framework. This means we can think of void loop as the main.

Declaring a function is how we would call the function in our loop. This is also known as function prototyping. This is one kind of declaring.

There is also declaring a function when making a library. In that case it is required to include the type. This is a very important thing to know. But that is a more advanced topic.

Back to our questions:

Wouldn't the functions be the same type as the setup and the loop?

1. No, some of the functions we create can be void. They are used for side effect. When we are not concerned about returning a value.

2. Other functions would return a value. In that case they would be the same type as the value returned. For when you are concerned about returning a value. There are many cases for this.

What type should the functions be?

> It would be the same type as the values it would return.
>
> The values would be stored in the variables we use. That means it would be the type of the variables.

We should now have the answer to what type our functions should be.

All our variables are type int. The name should be something that describes what the function does. Here we will name the first function sDots and the second oDashes.

We can now get into programming.

Adding Functions To The Program

Remember our short Algorithm.

// Create function for S

// Create function for O

```
27      /// S
28      //
29      // Create function for S
30      sDot();
31
32      // turn the LED on (HIGH is the v
33      digitalWrite(pin13, HIGH);
34      // wait for a second
35      delay(_1second);
36      // turn the LED off ( LOW is the
37      digitalWrite(pin13, LOW);
38      // wait for a second
```

Let's go ahead and declare(function prototype our function underneath the comments.. Then add them to our code.

The syntax for declaring(function prototyping is)

      <name> ( );

This will call the function we are creating in our loop..

Warning: If you try to verify it now you will get an error.

Error:

      morseSos.ino: In function 'void loop()':

Uno Programming digitalwrite   Ebonygeek45

morseSos:28: error: 'sDot' was not declared in this scope.

'sDot' was not declared in this scope.

You're not doing anything wrong. This is a case of, the code is not complete yet. We just declared the function. The compiler looked for the definition of the function and can't find it. That is because we have not made it yet.

'sDot' was not declared in this scope.

What does scope mean? You ask.

When calling a function, it must be declared. If we had defined it and we got that error, we would bring it into scope by declaring it.

We declared it here. Vice versa, it must be defined. We have not defined it. To bring sDot into Scope we have to define it.

So let's do that.

All it takes is to add the function.

```
<type> <name>( )
{
    // Instructions for the function
    // <code>
```

}

Would be:

    int sDot( )

    {

        // Instructions for the function

        // <code>

    }

We have it declared in the loop where we want the function to run.

Where does the definition of the function go? You ask.

It can go above the setup function or below the loop function.

Here we will put it below the loop function(after its closing curly bracket).

That will define the sDot function and bring it into scope.

```
122        delay(_4sbetweenLetters);
123
124 }
125
126 int sDot( )
127 {
128    // Instructions for the function
129    // <code>
130 }
131
132
```

Done compiling.
Global variables use 9 bytes (0%) of dy

No, it doesn't have the code in it yet.

But, now it will verify.

Let's define it.

Well, that is as simple as cutting our top block of S code and pasting it where "// code" is.

Make sure to copy all the code for the first S. That is for all the dots.

Uno Programming digitalwrite

```
morseSos | Arduino 1.8.2
File Edit Sketch Tools Help

morseSos
122
123  int sDot( )
124  {
125    // Instructions for the function
126      /// S
127    // .
128    // Create function for S
129    // turn the LED on (HIGH is the voltage level)
130    digitalWrite(pin13, HIGH);
131    // wait for a second
132    delay(_1second);
133    // turn the LED off ( LOW is the voltage level)
134    digitalWrite(pin13, LOW);
135    // wait for a second
136    delay(_1second);
137    // .
```

The sDot function defined by the S block of code.

Uno Programming digitalwrite

```
morse8os §
21
22    // the loop function runs over and over again forever
23  void loop()
24  {
25     /// S
26       // .
27         // Create function for S
28     sDot();
29
30     /// O
31       // -
32         // Create function for O
33
34           // turn the LED on (HIGH is the voltage level)
35     digitalWrite(pin13, HIGH);
36     delay(_3second);        // wait for a second
37           // turn the LED on ( LOW is the voltage level)
```

Back in our loop the first S should now have a call to the function.

Verify and upload the code and you should have your same SOS blinks.

Think you got it?

Go ahead and do O.

Then the remaining S.

Summary:

We have been through a lot together.

Uno Programming digitalwrite   Ebonygeek45

We went from pinning out the circuit to adding functions to our program and everything between.

Maybe even went through a whole bag of cookies.

Good job.

Are we returning the function?

Yes we are.

It is a good practice to return functions. Why not develop good habits from the start.

In the next chapter, we are going to improve and return our functions.

CHAPTER 6: FUNCTIONS AND VARIABLES

Were you able to add the functions and upload them correctly?

Let's get on the same page. Your code should now be:
****

// morseSos.ino

// Variable Block

// Stores pin number

const int pin13 = 13;

// Stores 1 second value

const int _1second = 1000;

// Stores 3 second value

const int _3second = 3000;

// Stores 4 sec value

const int _4sbetweenLetters = 4000;

// setup function runs once when you press reset or power the

Uno Programming digitalwrite   Ebonygeek45

board

void setup()

{

   // initialize digital pin 13 as an output.

   pinMode(pin13, OUTPUT);

}

// the loop function runs over and over again forever

void loop()

{

   /// S

   // .

   sDot();

   /// O

   // -

   oDash();

/// S

// .

sDot();

}

int sDot( )

{

// .

// turn the LED on (HIGH is the voltage level)

digitalWrite(pin13, HIGH);

// wait for a second

delay(_1second);

// turn the LED off (LOW is the voltage level)

digitalWrite(pin13, LOW);

// wait for a second

delay(_1second);

// .

```
// turn the LED on (HIGH is the voltage level)

digitalWrite(pin13, HIGH);

// wait for a second

delay(_1second);

// turn the LED off ( LOW is the voltage level)

digitalWrite(pin13, LOW);

// wait for a second

delay(_1second);

// turn the LED on (HIGH is the voltage level)

digitalWrite(pin13, HIGH);

// wait for a second

delay(_1second);

// turn the LED off ( LOW is the voltage level)

digitalWrite(pin13, LOW);

// wait for a second

delay(_1second);
```

```
    // delay 4 seconds between letters

    delay(_4sbetweenLetters);

}


int oDash( )

{


    // -


    // turn the LED on (HIGH is the voltage level)

    digitalWrite(pin13, HIGH);

    // wait for 3 seconds

    delay(_3second);

    // turn the LED off ( LOW is the voltage level)

    digitalWrite(pin13, LOW);
```

```
// wait for 3 seconds
delay(_3second);


// -


// turn the LED on (HIGH is the voltage level)
digitalWrite(pin13, HIGH);
// wait for 3 seconds
delay(_3second);
// turn the LED off ( LOW is the voltage level)
digitalWrite(pin13, LOW);
// wait for 3 seconds
delay(_3second);


// -


// turn the LED on (HIGH is the voltage level)
```

```
digitalWrite(pin13, HIGH);

// wait for 3 seconds

delay(_3second);

// turn the LED off ( LOW is the voltage level)

digitalWrite(pin13, LOW);

// wait for 3 seconds

delay(_3second);


// delay 4 seconds between letters

delay(_4sbetweenLetters);



}
```
****

OK, Functions and Variables is the name of this chapter. Don't worry. It is not going through it again. This chapter will show us putting them both together to improve the code even more.

First lets change the speed of our blinks. We now know what it is supposed to do. We can handle more speed right.

It is simple to change the speed of the blinks.

Uno Programming digitalwrite   Ebonygeek45

Remember since the values are stored in variables, we only have to change it there.

We will leave the name as is for now.

Change the variable values to:

text

    _1second, will be change to 300.

    _3second, will be change to 900

    _4sbetweenLetters, will be change to 1000

It should now look like below:

```
 9  // Variable Block
10  const int pin13 = 13;                  // Stores pin num
11  const int _1second = 300;              // Stores 1 secon
12  const int _3second = 900;              // Stores 3 secon
13  const int _4sbetweenLetters = 1000;    // Stor
```

Verify and upload it.

This should show how simple it is to change the values stored in variables.

It looks more like Morse code because it is not blinking so slow.

Passing Variables To Functions

Uno Programming digitalwrite   Ebonygeek45

Let the function work for you, not you work for the functions.

Passing variables to the function can do that.

It can be a part of Optimizing your code in a major way.

Again, we are going to work through an Algorithm.

This new Algorithm is called:

    Passed value to function Algorithm

Shall we continue on.

To pass a value to a function, you can use a value itself IE 1000.

But, if you remember storing the value in a Variable is more efficient. Revisit the chapter on variables if you are foggy on what was shown there.

Our functions are already built out.

We are going to be changing them a little.

We have two functions now, the sDot and oDash.

Recap

As you should know by now. The functions store the instructions(the code) for the function.

Like a variable you then use it where ever you would have typed out your code.

Uno Programming digitalwrite   Ebonygeek45

It is a way to tell the Arduino IDE.

> Stop right here and pass control to this function.

> Find this function and follow its instructions (Do the code).

> Then come back here and run the next line of code in the function.

Taking Functions Further

We are taking it a step further by passing a value to the function.

* So it would then tell the Arduino IDE.

> Stop right here and pass control to this function.

> Find this function with this variable and follow its instructions (Do the code).

> Use this variable whenever the instructions(code) tell you to.

> Then come back here and run the next line of code in the function.

What is the point? You ask.

The advantage comes back to Optimizing code:

1. It reduces the lines of code in the loop.

2. It allows you to reuse a function that has different values.

Instead of many functions that is the same except for the values. You can have one function that can use many different values.

There are many more advantages than the ones above, those are my favorites.

To pass values into a function:

   1. You should create or already have your function first.

   2. Then declare it in the loop where you need it. That is, for any function. This will call that function.

We just went through that process in the last chapter. If you have to go back to it, do so.

We have our functions set up like that now.

This is where things will change.

Let's take a look at the functions that make use of passed values the Arduino IDE provides for us.

We have been using them all the time.

Uno Programming digitalwrite   Ebonygeek45

```
   morseSos
18       // initialize digital pin 13 as an output.
19       pinMode(pin13, OUTPUT);
20   }
21
22       // the loop function runs over and over agai
23   void loop()
24   {
43
44   int sDot( )
45   {
46           // turn the LED on (HIGH is the voltag
47       digitalWrite(pin13, HIGH);
48       delay(_1second);          // wait for a second
49           // turn the LED    ( LOW is the voltag
50       digitalWrite(pin13, LOW);
51       delay(_1second);          // wait for a secon
```

Did you realize that "pinMode", "digitalWrite", and "delay" are functions with passed values?

They are already made for your convenience and to ease you into using the Arduino IDE.

Imagine if you had to figure out how to create and use those functions before you could even start to try the blink.

There are many other functions with passed values already created just waiting for when you need them.

Notice they follow the syntax for declaring a function(function prototype).

The syntax for declaring(function prototyping is)

<name> ( );

```
26      // .
27          // Create function for S
28      sDot();
29
30      /// o
31      // -
32          // Create function for O
33
34      oDash();
35
36
37      /// S
38      // .
39          // Create function for S
40      sDot();
```

The difference is they have the advantage of passed values in their parameters.

Our functions do not have the advantage of passed variables.....yet.

As you can see there is nothing in our parameters.

Let's change that.

Uno Programming digitalwrite   Ebonygeek45

The question arises.

What would you pass to the functions?

Take a look at your "sDot" and "oDash" functions and look for repeating code.

What is being repeated over and over?.

The answer should be digitalWrite and delay.

OK, those are the functions we are calling so they are fine.

Now look at what we added to the functions digitalWrite and delay?.

It's in the parameters isn't it?

It's our variables.

Go up to your variables.

We should have 4 variables.

    // Stores pin number

    const int pin13 = 13;

    // Stores 1 second value

    const int _1second = 300;

    // Stores 3 second value

Uno Programming digitalwrite   Ebonygeek45

const int _3second = 900;

// Stores 4 sec value

const int _4sbetweenLetters = 1000;

Look at the functions and you should see that each function uses three of our variables within them. We don't consider HIGH and LOW because the Arduino IDE has already set them up properly. It is our variables that are not set up to their maximum advantage, or allowing our functions to be reusable. Also, pin13 is not really a concern either but it doesn't hurt to consider it as well.

Let's Get To Work

Passed value to function Algorithm:

1. First of all since we changed the values of our variables. Rename them to describe the values.

    1. for 300 Milliseconds

        dwDot

    2. for 900 Milliseconds

        dwDash

    3. for 1 Second:

        dwBetween

2. We are going to also need a variable to pass to the

function.

1. This variable is just declared an int type, but not defined with a value.

2. passing1

3. Make changes to the function with;

    1. Any new names.

    2. To allow it to pass the variables

4. Verify and Upload.

5. Break open that bag of cookies to celebrate.

Got it? Good.

We will start with one of our Algorithm. The variable really should have been renamed when the value was changed to keep down confusion.

We should now have:

   // Stores pin number

   const int pin13 = 13;

   // Stores 300 Milliseconds value

   const int dwDot = 300;

   // Stores 900 Milliseconds value

Uno Programming digitalwrite	Ebonygeek45

const int dwDash = 900;

// Stores 1 Second value

const int dwBetween = 1000;

Simple enough.

Notice we are not declaring the new variable passing1 here.

Now down to our function.

We are going to use sDot for our example here.

```
44  //    int sDot()
45        int sDot(int passing1)
46      {
```

We declare our passing1 variable in the sDot function's parameters.

The syntax is:

<type> <functionName>(<type> variableName>)

{ Instructions for function using passed variable }

When passing a variable to a function you do not define it like the other variables we used. It is a placeholder for the

Uno Programming digitalwrite					Ebonygeek45

variables you will be passing to it. As in, it is just holding a place for something else. Notice I said the variables. So keep that in mind for later.

Now we are going to go into the function and add passing1 to all the delays but the last one. Remember the last one was for the delay between letters. At this point we will change its variable name to the new one we gave it:

    dwBetween

Our sDot function should now be:

int sDot(int passing1)

{

    // turn the LED on (HIGH is the voltage level)

    digitalWrite(pin13, HIGH);

    // wait for 300 milliseconds

    delay(passing1);

    // turn the LED off (LOW is the voltage level)

    digitalWrite(pin13, LOW);

    // wait for 300 milliseconds

    delay(passing1);

    // .

Uno Programming digitalwrite

// turn the LED on (HIGH is the voltage level)

digitalWrite(pin13, HIGH);

// wait for 300 milliseconds

delay(passing1);

// turn the LED off (LOW is the voltage level)

digitalWrite(pin13, LOW);

// wait for 300 millisecondsdelay(passing1);

// .

// turn the LED on (HIGH is the voltage level)

digitalWrite(pin13, HIGH);

// wait for 300 milliseconds

delay(passing1);

// turn the LED off (LOW is the voltage level)

digitalWrite(pin13, LOW);

// wait for 300 milliseconds

delay(passing1);

Uno Programming digitalwrite   Ebonygeek45

```
    // delay 1 seconds between letters

    delay(dwBetween);

}
```

Now we have got to go back up to the loop and add our variable we have defined with the value in the parameters for the sDot function.

```
7     /// S
8        // .
9           // Create function for S
0        sDot(dwDot);
```

Verify and upload.

The sDot function should be good. But you are going to have a bunch of other errors.

Because?

Well, I am just going to leave that for you as a challenge for now.

Go through the code and figure out what need to be corrected and why.

The oDash function needs to be done too. See if you can do it on your own.

Once all the errors and oDash is done come back and check to see if we agree. If it's over your head come and see what got you stumped.

There's more to come so hurry back.

Figuring Out Errors

You are going to run into errors all the time. A missing colon, bracket, something that was not changed, etc, etc, etc. That is the life of a programmer. Those little bugs like to keep you thinking.

When dealing with hardware(the Uno) you will have more mistakes. The wrong pin, cord not plugged in, damaged components, etc, etc, etc.

Believe it or not, that is part of the fun. The more you get into it, the more you will know how to quickly correct the problems.

Let's go over the error you may have run into in your task of passing the variable to the oDash function.

These errors are telling you that the variables in the oDash function "was not declared in this scope.

We saw a scope error before remember?

So what does that mean?

Well, it is also give you our old variable names. That should tell you that just maybe it is letting us know those variables are no longer in use.

What we need to do is change them.

When we set up the oDash function to pass variables that should solve that problem.

A lot of times in programming you get errors because you are not finished adding your programming. IE when declaring a function in the loop, and you get an error that it was not in scope because it is not defined yet.

Programming The oDash Function

Our updated code is below:
****

// morseSos.ino

// Variable Block

// Stores pin number

const int pin13 = 13;

// Stores 300 millisecond value

const int dwDot = 300;

// Stores 900 millisecond value

const int dwDash= 900;

// Stores 1 second value

const int dwBetween = 1000;

/*

setup function runs once when you press reset or power the board.

*/

Uno Programming digitalwrite   Ebonygeek45

void setup()

{

    // initialize digital pin 13 as an output.

    pinMode(pin13, OUTPUT);

}

// the loop function runs over and over again forever

void loop()

{

    /// S

    // .

    sDot(dwDot);

    /// O

    // -

    oDash(dwDash);

    // S

    // .

    sDot(dwDot);

}

```
int sDot(int passing1)
{
    // .
    // turn the LED on (HIGH is the voltage level)
    digitalWrite(pin13, HIGH);
    // wait 300 milliseconds
    delay(passing1);
    // turn the LED off( LOW is the voltage level)
    digitalWrite(pin13, LOW);
    // wait 300 milliseconds
    delay(passing1);
    // .
    // turn the LED on (HIGH is the voltage level)
    digitalWrite(pin13, HIGH);
    // wait 300 milliseconds
    delay(passing1);
```

Uno Programming digitalwrite

```
// turn the LED off ( LOW is the voltage level)
digitalWrite(pin13, LOW);
// wait 300 milliseconds
delay(passing1);
// .

// turn the LED on (HIGH is the voltage level)
digitalWrite(pin13, HIGH);
// wait 300 milliseconds
delay(passing1);
// turn the LED off ( LOW is the voltage level)
digitalWrite(pin13, LOW);
// wait 300 milliseconds
delay(passing1);


// delay 1 seconds between letters
delay(dwBetween);
}
```

```
int oDash(int passing1)
{
    // -
    // turn the LED on (HIGH is the voltage level)
    digitalWrite(pin13, HIGH);
    // wait 900 milliseconds
    delay(passing1);
    // turn the LED off ( LOW is the voltage level)
    digitalWrite(pin13, LOW);
    // wait 900 milliseconds
    delay(passing1);

    // -
    // turn the LED on (HIGH is the voltage level)
    digitalWrite(pin13, HIGH);
    // wait 900 milliseconds
```

Uno Programming digitalwrite

```
delay(passing1);

// turn the LED off ( LOW is the voltage level)
digitalWrite(pin13, LOW);

// wait 900 milliseconds
delay(passing1);

// -

// turn the LED on (HIGH is the voltage level)
digitalWrite(pin13, HIGH);

// wait 900 milliseconds
delay(passing1);

// turn the LED off ( LOW is the voltage level)
digitalWrite(pin13, LOW);

// wait 900 milliseconds
delay(passing1);


// delay 1 seconds between letters
delay(dwBetween);
```

}
****

Where's the cookies?, and on to the next chapter.

Summary:

This one is simple. But for some people passing variables into functions is tricky. The main thing to remember is that the variable declared in the function definition's parameters is a placeholder. It is holding the place for the variable that we use when calling the function.

Once you understand that you are good to go.

If you run into problems above your head, Google and YouTube are a lifesaver. Also, C++ forums. In these forums you can create a post with your problem and people will help most of the time. You may even get extra advise that you may or may not choose to use.

CHAPTER 7: MORE PASSING VARIABLES

So what about the other variables in the function?

We only passed one variable to the function, but there are two more isn't it? These can be passed too.

```
25    /// S
26     // .
27    sDot(dwDot, pin13, dwBetween);
28    /// O
29     // -
30
31    oDash(dwDash);
32
```

Passing Multiple Variables

We simply declare the other two variables in our function prototype in the loop.

Notice the error. We have seen this before. It is because I verify often. When the sDot function is defined for the other 2 variables, the errors will go away.

Go down to the sDot function and we are going to add two more placeholder variables to its parameters.

Remember to also change the second "S" underneath the "O".

When you add to the parameters of a function they are called

Uno Programming digitalwrite

arguments. So we now have three arguments in the parameters.

In the loop pin13 is declared as the second argument and dwBetween is the third.

All that need to be done now is to go into the function and change the arguments for digitalWrite and the last delay.

The sDot function is below as it should be now:

****

// int sDot()

int sDot(int passing1, int passing2, int passing3)

{

    // turn the LED on (HIGH is the voltage level)

    digitalWrite(passing2, HIGH);

    // wait 300 milliseconds

    delay(passing1);

    // turn the LED off (LOW is the voltage level)

    digitalWrite(passing2, LOW);

    // wait 300 milliseconds

    delay(passing1);

// .

// turn the LED on (HIGH is the voltage level)

digitalWrite(passing2, HIGH);

// wait 300 milliseconds

delay(passing1);

// turn the LED off (LOW is the voltage level)

digitalWrite(passing2, LOW);

// wait 300 milliseconds

delay(passing1);

// .

// turn the LED on (HIGH is the voltage level)

digitalWrite(passing2, HIGH);

// wait 300 milliseconds

delay(passing1);

// turn the LED off (LOW is the voltage level)

digitalWrite(passing2, LOW);

// wait 300 milliseconds

delay(passing1);

// delay 1 seconds between letters

delay(passing3);

}
****

One function?

You have probably noticed that there is only one thing that is different about the sDot and oDash function.

What if I told you that we can use the same function for both?

Do you know how to make that possible?

Instead of setting up the oDash function the same way as the sDot function. Try to change sDot to work for both the functions.

Think you can handle it?

Go ahead and try.

Well, I had a cup of coffee and watched a movie. I love superheroes.

How did you do?

Uno Programming digitalwrite   Ebonygeek45

Let's go through it together. Going through my algorithm. That is the first thing you should have done is an algorithm.

Change the name of the sDot function to match what it will do.

In the loop:

    // S

    // .

    sDot(dwDot, pin13, dwBetween);

Changed to:

    // S

    // .

    theBlinks(dwDot, pin13, dwBetween);

It is still going to be S but the function name is changed because the function is going to be used by O too. Remember to change both the top and bottom S.

Going down to the sDot function to change it. These changes are going to make it reusable.

> Change the name to theBlinks.

    int sDot(int passing1, int passing2, int passing3)

Uno Programming digitalwrite   Ebonygeek45

Changed to:

    int theBlinks(int passing1, int passing2, int passing3)

That change is what makes the magic happen. Remember the variables in the parameters(Arguments) are place holders for whatever variables you want to use when you call the function.

That is the only changes we need to make. The functions, instructions inside the curly brackets is already set up to work for O too. The passed variables make this possible.

Now go back up to the loop. We are now going to change the O function prototype to use what is now theBlink function.

In the loop:

    // O

    // -

    oDash(dwDash);

Changed to:

    // O

    // -

    theBlinks(dwDash, pin13, dwBetween);

Verify and Upload

Does your led blink the SOS?

Good.

If you didn't do an algorithm, maybe later you will see the use for them.

Sometimes you don't know what steps you will do. But after you figure it out it is a good idea to do an algorithm.

Why? You ask.

For future reference. That way you will know how to do something the next time you come to it. It is a good thing to have a folder full of algorithms. That way you can look up things you may have forgotten how to do.

Back to it.

See how the function is now reusable. Our code has been further optimized.

Go ahead and delete the oDash function outside of the loop. It is no longer needed.

Checking The Sketch

Remember that sketch is how the Arduino folks refer to their .ino file.

To make sure we are on the same page, the sketch is below.
****

// morseSos.ino

// Variable Block

Uno Programming digitalwrite

```
// Stores pin number
const int pin13 = 13;
// Stores 300 millisecond value
const int dwDot = 300;
// Stores 900 millisecond value
const int dwDash= 900;
// Stores 1 second value
const int dwBetween = 1000;


/*  setup function runs once when you press reset or power the board */
void setup()
{
    // initialize digital pin 13 as an output.
    pinMode(pin13, OUTPUT);
}
// the loop function runs over and over again forever
void loop()
```

```
{
    // S
    //
    theBlinks(dwDot, pin13, dwBetween);
    // O
    // -
    theBlinks(dwDash, pin13, dwBetween);
    // S
    // .
    theBlinks(dwDot, pin13, dwBetween);
}
int theBlinks(int passing1, int passing2, int passing3)
{
    // turn the LED on (HIGH is the voltage level)
    digitalWrite(pin13, HIGH);
    // wait ? milliseconds
    delay(passing1);
```

Uno Programming digitalwrite

```
// turn the LED off ( LOW is the voltage level)
digitalWrite(pin13, LOW);
// wait ? milliseconds
delay(passing1);
// .
// turn the LED on (HIGH is the voltage level)
digitalWrite(pin13, HIGH);
// wait ? milliseconds
delay(passing1);
// turn the LED off ( LOW is the voltage level)
digitalWrite(pin13, LOW);
// wait ? milliseconds
delay(passing1);
// .
// turn the LED on (HIGH is the voltage level)
digitalWrite(pin13, HIGH);
// wait ? milliseconds
```

delay(passing1);

// turn the LED off (LOW is the voltage level)

digitalWrite(pin13, LOW);

// wait ? milliseconds

delay(passing1);

// delay 1 seconds between letters

delay(dwBetween);

return;

}
****

Summary:

We have gone from two functions to blink s and o for the SOS blinks to one Function.

First, we passed one variable, then three.

You can pass as many variables as needed for a function.

Variables are very handy and later we will see what else they can do.

But, when passing more than one variable you can only return one value. This is something that stumped me for a long time.

What are you talking about? You ask.

Returning values are optional. Remember void is for when you are not concerned about returning the value(as in there is really no point). But it is the correct way to code in a lot of cases. As you get into more advanced programming it is required. Simply return which means returning control of the function back to the loop.

On to the next chapter...

On your marks...get set...go to the next chapter....>>>

CHAPTER 8: BEHAVIORS AND ATTRIBUTES

We are still dealing with functions. It is a big subject. If you know how to work with them the right way, your programming will go smoother.

The right way will vary depending on who you discuss programming with. When you get more into programming you will develop your own style.

What I am going through here is my style and the way I understand it. It always works for me and the way I see it. If it aint broke, don't fix it.

Behaviors and Attributes are a way of organizing your code. The more organized you can be, the better. It is a way of planning your code. It also can help you understand what is needed for your programs.

Behaviors

Behavior is just that. How your function will behave. What it will do.

Behaviors are the functions(methods) of your programming. A way of thinking about it is English grammar. It is the verbs of your programming. It is the actions of the program.

Look at the functions void setup and void loop.

Both of those are behaviors. Their instructions are how the functions will behave. What they will do.

The setup function is for code that need to only run once. That is its behavior. This is important because you may create a function that you only want to run once.

If you put it in the loop function, it will behave as a loop does and run over and over till infinity.... you get the idea. Looping is its behavior.

Attributes

Attributes can change.

They are the variables that you use in your functions. They describe the behavior. I would say think of them like adjectives. They describe the state of an object(I know we haven't gotten into objects yet, but it is coming in the next book). For now I would say they describe the state of our functions.

For our pin13 variable it can change to whatever pin. It describes what pin will be used.

Same for theBlinks, the variables in them can change. For example the way we changed our sdot from 1000 to 300, and odash from 3000 to 900. They describe how long the blinks should be.

If you think of children in a real life scenario

Their attributes could be:

    name, age, gender, weight, etc

Their Behaviors would be:

play, misbehave, study, sleep, etc

You see how that works together.

Your attributes would be your variables. You can already see there are two different types for the ones we have listed.

The behaviors would be the functions. Their types would depend on the type of the attribute they use.

It is a very simple concept. Thinking about it like that will help you in being able to plot our your programs.

Especially in more advanced topics that hopefully I will help you through.

Now back to our code, let's see how we would apply the above theory.

Putting it all together

We have one function for our program.

We have optimized it so it is handling both "s" and "o".

We are not going to do any coding. This is an easy chapter for you.

We are adding to the comments to show your Behaviors and Attributes. It is really simple.

Checking The Code

Uno Programming digitalwrite

```
/*

The color will highlight what it is.

Attributes, Behaviors, Arduino IDE Attributes, Arduino IDE Behaviors.

*/

// Attributes

// Stores pin number

const int pin13 = 13;

// Stores 300 millisecond value

const int dwDot = 300;

// Stores 900 millisecond value

const int dwDash= 900;

// Stores 1 second value

const int dwBetween = 1000;


/*
```

Uno Programming digitalwrite   Ebonygeek45

setup function runs once when you press reset or power the board

*/

void setup()

{

   // initialize digital pin 13 as an output.

   pinMode(pin13, OUTPUT);

}

// the loop function runs over and over again forever

void loop()

{

   // S

   // .

   theBlinks(dwDot, pin13, dwBetween);

   // O

   // -

Uno Programming digitalwrite

```
    theBlinks(dwDash, pin13, dwBetween);

    /// S

    // .

    theBlinks(dwDot, pin13, dwBetween);
}


// Behaviors

int theBlinks(int passing1, int passing2, int passing3)
{
    // turn the LED on (HIGH is the voltage level)
    digitalWrite(pin13, HIGH);
    // wait ? milliseconds
    delay(passing1);
    // turn the LED off ( LOW is the voltage level)
    digitalWrite(pin13, LOW);
```

// wait ? milliseconds

delay(passing1);

// turn the LED on (HIGH is the voltage level)

digitalWrite(pin13, HIGH);

// wait ? milliseconds

delay(passing1);

// turn the LED off (LOW is the voltage level)

digitalWrite(pin13, LOW);

// wait ? milliseconds

delay(passing1);

// turn the LED on (HIGH is the voltage level)

digitalWrite(pin13, HIGH);

// wait ? milliseconds

delay(passing1);

// turn the LED on (LOW is the voltage level)

Uno Programming digitalwrite   Ebonygeek45

    digitalWrite(pin13, LOW);

    // wait ? milliseconds

    delay(passing1);

    // delay 1 seconds between letters

    delay(dwBetween);

}
****

Summary:

This chapter is meant to give you an understanding of how the program is set up.

It is also to show you how you can use Behaviors and Attributes to help you plot out your program when doing an algorithm.

As you can see the Behaviors(Functions) and Attributes(Variables) cover most of the coding in the sketch.

No coding in this chapter, but this chapter is very important as far as planning your code.

Nuff said. On to For Loops.

CHAPTER 9: FOR LOOP

Looking through the code we left off with in the last chapter. There is a lot of repeating code.

We have tried other things to manage it:

> Added Variables

> Created functions.

> Passed variables to those functions.

> Shown Attributes and Behaviors.

Let's try loops.

Loops are used to repeat blocks of code.

Seems like just what we need for our function.

We have been coding the digitalWrite dots 3 times for the S and the same thing for the dash.

We can create a loop to do this for us.

The For Loop

We will use a for loop.

// Syntax for the for loop.

```
for(<int><varName> = <startNumber>; <varName> < <endNumber>; <varName>++)

{
        Code to loop here
}
```

Breaking it down:

The For Loop:

Syntax is:

```
for (int e = 0; e < 3; e++)
{
        // Code to loop here
}
```

type of Loop	var start Loop	var times to Loop	var add each time
for (	int e = 0;	e < 3;	e++)

```
{
        Instruction code to for Loop
}
```

Uno Programming digitalwrite   Ebonygeek45

A Little more Breakdown.

The Syntax and break down

for(check if condition is true)
{
    Run this code
}

(int e = 0;   )

e is a variable. It will start the loop at 0. It can be any letter or name used in these parameters.

(int e = 0;   e < 3;   )

As long as e is less than 3. This number can change per how many times it need to loop.

(int e = 0;   e < 3;   e++)

Increment(add) 1 if less then the number stored in e.

Let's plot out an algorithm shall we.

Our digitalWrite's and delay's are what we are looking at now.

For Loop Algorithm for theBlinks.

1. <type> will be int.

2. <varName> will be e.

    1. It can be any variable name you want. Here we are keeping it simple. It is only going to be used in this for loop.

    2. It should not need to be declared because it is being declared in the for loop parameters.

Uno Programming digitalwrite

3. <startNumber> will be 0.

   1. StartNumber should always be 0.

   2. It will count up from 0 to the endNumber.

4. <endNumber> will be 3.

   1. 3 is the endNumber because that is how many times we will need it to loop.

   2. Remember, we have 3 identical blocks of code for the dots and the dashes. If you don't see that take a good look at the code for "theBlinks".

   3. We will leave the last delay as it is because we need that delay between the blinks. It will not be in the for Loop.

Your final test.

Try it on your own and see what you come out with.

Don't over-think it.

When you are done, come back here for the solution below.

Checking The Sketch
****

// morseSos.ino

```
// Attributes
// Stores pin number
const int pin13 = 13;
// Stores 300 millisecond value
const int dwDot = 300;
// Stores 900 millisecond value
const int dwDash = 900;
// Stores 1 second value
const int dwBetween = 1000;

/*
setup function runs once when you press reset or power the board
*/
void setup()
{
    // initialize digital pin 13 as an output.
    pinMode(pin13, OUTPUT);
}

// the loop function runs over and over again forever
void loop()
```

{

    // S

    // .

    theBlinks(dwDot, pin13, dwBetween);

    // O

    // -

    theBlinks(dwDash, pin13, dwBetween);

    // S

    // .

    theBlinks(dwDot, pin13, dwBetween);
}

// Behaviors

int theBlinks(int passing1, int passing2, int passing3)

{

    for (int e = 0; e < 3; e++)

```
    {
        // turn the LED on (HIGH is the voltage level)

        digitalWrite(pin13, HIGH);

        // wait ? milliseconds

        delay(passing1);

        // turn the LED off( LOW is the voltage level)

        digitalWrite(pin13, LOW);

        // wait ? milliseconds

        delay(passing1);

    }


    // delay 1 seconds between letters

    delay(dwBetween);
}
```
****

Summary:

That was a short chapter, and it shortened our code. Loops can be very helpful for repeating code.

Uno Programming digitalwriteEbonygeek45

There are three kinds of loops:

    for (Most useful loop, we used this in our code.)

    while

    do...while

They all have their separate uses. They can be extremely helpful. People find the most interesting way to make these loops work for them. If you find them confusing, don't give up. Once you figure it out it is going to save you a lot of hassle.

Did you find it difficult to figure out?

Most of the time programmers over-think the coding. In some cases that is good. In some cases not so good.

Take it in small chunks.

Especially when going through someone else's code.

Whatever you do, don't give up.

Experiment and research..research...research.

CHAPTER 10: IN CLOSING

The very minimal electronic components were chosen to show results for the coding in this book. That was done intentionally. When starting out I didn't have or could not afford LED's. I used old Christmas lights. LED's can be salvaged off old electronics. Most of the time when first starting out you don't have a lot of components to work with.

Most of the time when geeks, nerds, programmers, etc put out a book they forget the new comers. The books are hard to follow for someone just beginning. The tech speak confusing to them. All attempts were made to make this book as easy to follow as possible. The focus to encourage to keep going forward. Not to frustrate to the point someone gives up. I hope I was successful in making this an enjoyable experience.

This books focus is on the programming behind the components. Look up the official Arduino site:
www.arduino.cc

Check out their site for some very interesting things, including building your own Uno. They also have some useful tutorials and go over the basics of coding as well.

The C++ programming world is full of ways to optimize and

improve the code. You will learn different and better ways to program as you develop your craft. The very basics were explained here and there is still more.

On the programming side.
- If Statement
- Switch and Case
- Arrays
- etc...etc...

On the Uno side.
- Serial Monitor
- Buttons
- Potentiometers
- RGB LED's
- etc...etc....

This book is an experiment to see if there is a demand for the next one in the series I am planning to start.

This will continue on with the programming side, maybe add some more components. So we will still be stuck with our SOS. But because of the quick results, it is a good setup for going over programming.

We will go over:
1. Setting up the Serial Monitor to communicate
2. If statement

3.   Switch and Case
4.   Arrays
5.   Classes
6.   Libraries
7.   etc...etc...

You may also want to take a look at my YouTube channel.

There are videos on getting set up with the Eclipse Arduino IDE. There are other videos you may be interested in.

Just pull up YouTube and search for Ebonygeek45.

If you find errors in this book, or if you find that you can not get the code to work, email me

ebonynerd45@gmail.com

Soon to come on my YouTube Channel: New videos and more complete series. Maybe based on this book and any other future ones.
That depends on the demand for this book.

If there are publishers that would like to publish future books

by myself, I am very open to speaking with you.

My new website is in development and should be live soon.

Ebonygeek45 is also looking for supporters and sponsors to

try to get out more material in a more timely bases, and to

expand production crew from just me to others who can help.

With everything going on a lot is changing and being redone.

Very soon the new website will be set in place to give any up

Uno Programming digitalwrite   Ebonygeek45

to date news and information.

It has been a treat going on this journey with you.

Happy tinkering, Happy coding.

Today is a good day to learn something.

Ebonygeek45

ABOUT THE AUTHOR

I live in Gary Indiana. Where I became the proud owner of my first home. With that first home I have also learned I am responsible for each expense and repair. Gary Indiana is not a good place to find gainful employment. Every extra penny I have goes to getting the things I need to share my knowledge. I am self taught and the internet is my classroom where I learn. My videos are on YouTube under Ebonygeek45. If I can and do. Then you can to.

www.ingramcontent.com/pod-product-compliance
Lightning Source LLC
Chambersburg PA
CBHW071440180526
45170CB00001B/393